CROP SALES

		Corn			Soybeans					
Date	What and to Whom Sold	Bushels	Amount		Bushels	Amount			Amount	
			$			$			$	
	Totals		$			$			$	

CROP SALES

		Hay and Other Forages			Fruits and Vegetables			Other Crops		
Date	What and to Whom Sold	Quantity	Amount		Quantity	Amount		Quantity	Amount	
			$			$			$	
Totals			$			$			$	

HOG SALES

Date	What and to Whom Sold	No.	Weight	Amount if Raised		Amount if Purchased		Original Cost if Purchased		Qualify for Capital Gains?
				$		$		$		
	Totals			$		$		$		

Enter gross sales. Record deductions under Farm Expenses. Separate livestock purchased or raised on the farm. Enter the original cost of purchased animals. Check whether sales qualify for capital gains for income tax reporting (breeding stock must be owned at least 12 months).

BEEF CATTLE SALES

Date	What and to Whom Sold	No.	Weight	Amount if Raised		Amount if Purchased		Original Cost if Purchased		Qualify for Capital Gains?
				$		$		$		
	Totals			$		$		$		

Enter gross sales. Record deductions under Farm Expenses. Separate livestock purchased or raised on the farm. Enter the original cost of purchased animals. Check whether sales qualify for capital gains for income tax reporting (breeding stock must be owned at least 24 months).

DAIRY CATTLE SALES

Date	What and to Whom Sold	No.	Weight	Amount if Raised		Amount if Purchased		Original Cost if Purchased		Qualify for Capital Gains?
				$		$		$		
	Totals			$				$		

er gross sales. Record deductions under Farm Expenses. Separate livestock purchased or raised on the farm. Enter the original cost of purchased nals. Check whether sales qualify for capital gains for income tax reporting (breeding stock must be owned at least 24 months).

OTHER LIVESTOCK SALES

Date	What and to Whom Sold	No.	Weight	Amount if Raised		Amount if Purchased		Original Cost if Purchased		Qualify for Capital Gains?
				$		$		$		
	Totals			$		$		$		

Enter gross sales. Record deductions under Farm Expenses. Separate livestock purchased or raised on the farm. Enter the original cost of purchased animals. Check whether sales qualify for capital gains for income tax reporting. Other livestock includes sheep, goats, horses, poultry, etc.

DAIRY PRODUCT SALES

Date	Quantity	Price	Amount
		$	$
Totals			$

Enter gross sales. Record deductions under Farm Expenses.

OTHER LIVESTOCK PRODUCT SALES

Date	Quantity	Price	Amount
		$	$
	Totals		$

FARM PRODUCTS CONSUMED AT HOME

Date	Quantity	Type	Value
			$
	Totals		$

OTHER FARM INCOME

Date	Description	Agricultural Program Payments		Custom Hire		Crop Insurance Proceeds		Fuel and Other Refunds		Other Income	
		$		$		$		$		$	
	Totals	$		$		$		$		$	

United States Department of Agriculture marketing loans can be entered here or under Loans and Loan Payments, but not both.

MACHINERY, LAND, and IMPROVEMENTS PURCHASED or CONSTRUCTED

Date	Description	Amount		Date	Description	Amount	
		$				$	
Totals		$		Totals		$	

For items acquired by trading a previously owned asset, include only the dollars paid to complete the trade.
Include total costs for materials, labor, and services for improvements constructed on the farm.
Enter beginning basis value on income tax depreciation schedule.

MACHINERY, LAND, and IMPROVEMENTS SOLD or DISPOSED OF

Date	Description	Amount		Date	Description	Amount	
		$				$	
Totals		$		Totals		$	

ote sales expenses here. Delete items from income tax depreciation schedule.

LIVESTOCK PURCHASES

Date	What and from Whom Purchased	Hogs				Beef Cattle			
		No.	Weight	Amount		No.	Weight	Amount	
				$				$	
	Totals			$				$	

LIVESTOCK PURCHASES

		Dairy Cattle				Other Livestock			
Date	What and from Whom Purchased	No.	Weight	Amount		No.	Weight	Amount	
				$				$	
Totals				$				$	

Other livestock includes sheep, goats, horses, poultry, etc.

FARM EXPENSES

Date	Description	Chemicals	Conservation Expenses	Custom Hire	Employee Benefits	Feed Purchased
	Specify the item purchased	Insecticides Herbicides Innoculants Others	Nondepreciable earthwork	Machinery hiring Services hired	Insurance Meals Housing Clothing	Grain Hay Supplements Salt, Minerals
	Totals	$	$	$	$	$

Carry totals forward to page 32.

FARM EXPENSES

Date	Description	Fertilizers and Lime		Freight and Trucking		Fuel and Lubricants		Insurance		Labor Hired	
	Specify the item purchased	Commercial fertilizer Manure Lime		Grain Livestock Supplies		For vehicles Drying fuel Lubricants		Property Crop Liability		Wages not recorded on page 35	
Totals		$		$		$		$		$	

Carry totals forward to page 33.

FARM EXPENSES

Date	Description	Pension Plans	Rent and Lease Payments	Repairs and Maintenance	Seeds and Plants	Storage
	Specify the item purchased	For employees	Land Machinery Livestock Buildings Vehicles	Parts Labor Services	Seeds Plants Technical fees	Warehouse Rent
	Totals	$	$	$	$	$

Carry totals forward to page 34.

FARM EXPENSES

Date	Description	Supplies		Farm Taxes		Utilities		Veterinary and Breeding		Others	
	Specify the item purchased	Tools Small equipment Office items		Personal Property Real estate Licenses		Electricity Telephone Water		Services Semen Drugs		All other farm expenses	
	Totals	$		$		$		$		$	

Carry totals forward to page 35.

FARM EXPENSES

Date	Description	Chemicals	Conservation Expenses	Custom Hire	Employee Benefits	Feed Purchased
	Specify the item purchased	Insecticides Herbicides Innoculants Others	Nondepreciable earthwork	Machinery hiring Services hired	Insurance Meals Housing Clothing	Grain Hay Supplements Salt, Minerals
	Totals	$	$	$	$	$

Carry totals forward to page 32.

FARM EXPENSES

Date	Description	Fertilizers and Lime		Freight and Trucking		Fuel and Lubricants		Insurance		Labor Hired	
	Specify the item purchased	Commercial fertilizer Manure Lime		Grain Livestock Supplies		For vehicles Drying fuel Lubricants		Property Crop Liability		Wages not recorded on page 35	
Totals		$		$		$		$		$	

Carry totals forward to page 33.

FARM EXPENSES

Date	Description	Pension Plans	Rent and Lease Payments	Repairs and Maintenance	Seeds and Plants	Storage
	Specify the item purchased	For employees	Land Machinery Livestock Buildings Vehicles	Parts Labor Services	Seeds Plants Technical fees	Warehouse Rent
	Totals	$	$	$	$	$

Carry totals forward to page 34.

FARM EXPENSES

Date	Description	Supplies		Farm Taxes		Utilities		Veterinary and Breeding		Others	
	Specify the item purchased	Tools Small equipment Office items		Personal Property Real estate Licenses		Electricity Telephone Water		Services Semen Drugs		All other farm expenses	
	Totals	$		$		$		$		$	

Carry totals forward to page 35.

FARM EXPENSES

Date	Description	Chemicals	Conservation Expenses	Custom Hire	Employee Benefits	Feed Purchased
	Specify the item purchased	Insecticides Herbicides Innoculants Others	Nondepreciable earthwork	Machinery hiring Services hired	Insurance Meals Housing Clothing	Grain Hay Supplements Salt, Minerals
	Totals	$	$	$	$	$

Carry totals forward to page 32.

FARM EXPENSES

Date	Description	Fertilizers and Lime	Freight and Trucking	Fuel and Lubricants	Insurance	Labor Hired
	Specify the item purchased	Commercial fertilizer Manure Lime	Grain Livestock Supplies	For vehicles Drying fuel Lubricants	Property Crop Liability	Wages not recorded on page 35
Totals		$	$	$	$	$

Carry totals forward to page 33.

FARM EXPENSES

Date	Description	Pension Plans	Rent and Lease Payments	Repairs and Maintenance	Seeds and Plants	Storage
	Specify the item purchased	For employees	Land Machinery Livestock Buildings Vehicles	Parts Labor Services	Seeds Plants Technical fees	Warehouse Rent
	Totals	$	$	$	$	$

Carry totals forward to page 34.

FARM EXPENSES

Date	Description	Supplies	Farm Taxes	Utilities	Veterinary and Breeding	Others
	Specify the item purchased	Tools Small equipment Office items	Personal Property Real estate Licenses	Electricity Telephone Water	Services Semen Drugs	All other farm expenses
Totals		$	$	$	$	$

Carry totals forward to page 35.

FARM EXPENSES

Date	Description	Chemicals	Conservation Expenses	Custom Hire	Employee Benefits	Feed Purchased
	Specify the item purchased	Insecticides Herbicides Innoculants Others	Nondepreciable earthwork	Machinery hiring Services hired	Insurance Meals Housing Clothing	Grain Hay Supplements Salt, Minerals
	Totals	$	$	$	$	$

Carry totals forward to page 32.

FARM EXPENSES

Date	Description	Fertilizers and Lime		Freight and Trucking		Fuel and Lubricants		Insurance		Labor Hired	
	Specify the item purchased	Commercial fertilizer Manure Lime		Grain Livestock Supplies		For vehicles Drying fuel Lubricants		Property Crop Liability		Wages not recorded on page 35	
	Totals	$		$		$		$		$	

Carry totals forward to page 33.

FARM EXPENSES

Date	Description	Pension Plans		Rent and Lease Payments		Repairs and Maintenance		Seeds and Plants		Storage	
	Specify the item purchased	For employees		Land Machinery Livestock Buildings Vehicles		Parts Labor Services		Seeds Plants Technical fees		Warehouse Rent	
	Totals	$		$		$		$		$	

Carry totals forward to page 34.

FARM EXPENSES

Date	Description	Supplies		Farm Taxes		Utilities		Veterinary and Breeding		Others	
	Specify the item purchased	Tools Small equipment Office items		Personal Property Real estate Licenses		Electricity Telephone Water		Services Semen Drugs		All other farm expenses	
	Totals	$		$		$		$		$	

Carry totals forward to page 35.

FARM EXPENSES

Date	Description	Chemicals	Conservation Expenses	Custom Hire	Employee Benefits	Feed Purchased
	Specify the item purchased	Insecticides Herbicides Innoculants Others	Nondepreciable earthwork	Machinery hiring Services hired	Insurance Meals Housing Clothing	Grain Hay Supplements Salt, Minerals
	Totals	$	$	$	$	$

Carry totals forward to page 32.

FARM EXPENSES

Date	Description	Fertilizers and Lime	Freight and Trucking	Fuel and Lubricants	Insurance	Labor Hired
	Specify the item purchased	Commercial fertilizer Manure Lime	Grain Livestock Supplies	For vehicles Drying fuel Lubricants	Property Crop Liability	Wages not recorded on page 35
	Totals	$	$	$	$	$

Carry totals forward to page 33.

FARM EXPENSES

Date	Description	Pension Plans		Rent and Lease Payments		Repairs and Maintenance		Seeds and Plants		Storage	
	Specify the item purchased	For employees		Land Machinery Livestock Buildings Vehicles		Parts Labor Services		Seeds Plants Technical fees		Warehouse Rent	
	Totals	$		$		$		$		$	

Carry totals forward to page 34.

FARM EXPENSES

Date	Description	Supplies		Farm Taxes		Utilities		Veterinary and Breeding		Others	
	Specify the item purchased	Tools Small equipment Office items		Personal Property Real estate Licenses		Electricity Telephone Water		Services Semen Drugs		All other farm expenses	
	Totals	$		$		$		$		$	

Carry totals forward to page 35.

FARM EXPENSES

Date	Description	Chemicals		Conservation Expense		Custom Hire		Employee Benefits		Feed Purchased	
	Totals this page										
	Totals from page 12										
	Totals from page 16										
	Totals from page 20										
	Totals from page 24										
	Totals from page 28										
	Totals for the year	$		$		$		$		$	

FARM EXPENSES

Date	Description	Fertilizers and Lime	Freight and Trucking	Fuel and Lubricants	Insurance	Labor Hired
	Totals this page					
	Totals from page 13					
	Totals from page 17					
	Totals from page 21					
	Totals from page 25					
	Totals from page 29					
	Totals for the year	$	$	$	$	$

FARM EXPENSES

Date	Description	Pension plans		Rent and Lease Payments		Repairs and Maintenance		Seeds and Plants		Storage	
	Totals this page										
	Totals from page 14										
	Totals from page 18										
	Totals from page 22										
	Totals from page 26										
	Totals from page 30										
	Totals for the year	$		$		$		$		$	

FARM EXPENSES

Date	Description	Supplies		Farm Taxes		Utilities		Veterinary and Breeding		Others	
	Totals this page										
	Totals from page 15										
	Totals from page 19										
	Totals from page 23										
	Totals from page 27										
	Totals from page 31										
	Totals for the year	$		$		$		$		$	

CAR and TRUCK EXPENSES

Vehicle ______________________ Vehicle ______________________ Vehicle ______________________

Date	Description	Amount	Date	Description	Amount	Date	Description	Amount
		$			$			$
Total		$	Total		$	Total		$
Farm share	_____ %	$	Farm share	_____ %	$	Farm share	_____ %	$

Multiply the total for each column by the percent allocated to farm use.

RECORD of TAXABLE WAGE PAYMENTS and INCOME and SOCIAL SECURITY (FICA) TAX DEDUCTIONS

(This is a memorandum record for use in accumulating Social Security information for reporting purposes.)

Employee Name ______________________

Address ______________________

Social Security Number ______________________

Date	Taxable Earnings	Income Tax Deducted	FICA Tax Deducted	Net Amount Paid
	$	$	$	$
Total				

Employee Name ______________________

Address ______________________

Social Security Number ______________________

Date	Taxable Earnings	Income Tax Deducted	FICA Tax Deducted	Net Amount Paid
	$	$	$	$
Total				

Employee Name ______________________

Address ______________________

Social Security Number ______________________

Date	Taxable Earnings	Income Tax Deducted	FICA Tax Deducted	Net Amount Paid
	$	$	$	$
Total				

Employee Name ______________________

Address ______________________

Social Security Number ______________________

Date	Taxable Earnings	Income Tax Deducted	FICA Tax Deducted	Net Amount Paid
	$	$	$	$
Total				

LOANS RECEIVED

Date	From Whom	Purpose	Amount	Interest Rate	Loan Fees
			$	%	$
Total			$		$

Loan fees must be prorated over the repayment period of the loan. Carry total to page 40, other expenses.

INTEREST and PRINCIPAL PAYMENTS on NOTES, MORTGAGES, and CONTRACTS

Date	To Whom Paid	Purpose	Total Payment	Interest	Principal
			$	$	$
Total			$	$	$

Divide total loan payments into principal and interest.

NONFARM EXPENSES for TAX RECORDS

Date		Amount		Date		Amount	
	Nonfarm Taxes	$			Gifts to Charity	$	
	Medical and Dental				Interest Paid		
	Job Expenses						
	Other						

SUMMARY of FARM INCOME (for IRS Schedule F)

Sale of Crops	Page No.	Quantity	Amount	Other Farm Income	Page No.	Amount
Corn	1		$	Agricultural program payments	8	$
Soybeans	1			Custom hire	8	
Oats, wheat, small grains	1			Crop insurance proceeds	8	
Hay and other forages	2			Fuel and other refunds	8	
Fruits and vegetables	2			Other income	8	
Other crops	2			Cooperative distributions paid	39	
Total for crops			$	Total for other farm income		$

SALES of LIVESTOCK and LIVESTOCK PRODUCTS

Sales of Livestock and Livestock Products Raised	Page No.	Raised		Purchased				Amount Qualifying for Capital Gains
		Weight	Amount	No.	Weight	Amount	Original Cost	
Hogs	3		$			$	$	$
Beef cattle	4							
Dairy cattle	5							
Other livestock	6							
Dairy products	7							
Other livestock products	7							
Totals for livestock			$			$	$	$

SUMMARY of EXPENSES (for IRS Schedule F)

Description	Page No.	Amount	Description	Page No.	Amount
Car and truck expenses	34	$	Labor hired	29,35	$
Chemicals	28		Pension plans	30	
Conservation expenses	28		Rents and leases for machinery or vehicles	30	
Custom hire	28		Rents and leases for land or animals	30	
Employee benefits	28		Repairs and maintenance	30	
Feed purchased	28		Seeds and plants	30	
Fertilizer and lime	29		Storage	30	
Freight and trucking	29		Supplies	31	
Gasoline, fuel, and oil	29		Taxes	31	
Insurance	29		Utilities	31	
Interest—mortgage	36		Veterinary, breeding	31	
Interest—other	36		Other expenses	31	
			Total expenses		$

SUMMARY of ECONOMIC DEPRECIATION—Not for Use with Income Tax Return

	a	b	c	d	e	f	g
Items	Remaining Value from Previous Year	Purchases and New Construction (page 9)	Sales and Disposal (page 9)	Adjusted Value a + b - c	Depre-ciation Rate %	Depreciation This Year d x e	End of Year Remaining Value d - f
Machinery and equipment	$	$	$	$		$	$
Vehicles (farm share)							
Other intermediate assets							
Buildings							
Improvements							
Land							
Other long-term assets							
Total depreciation						$	$

Suggested depreciation rates are 10% for machinery, equipment, and vehicles and 5% for buildings and improvements. Land is not depreciated.

Round values to the nearest whole dollar.

SUMMARY of COOPERATIVE DISTRIBUTIONS

	a	b	c	d
Source	Accumulated Value from Previous Year	New Distributions Authorized	Distributions Paid Out	New Accumulated Value a + b - c
Totals				

SUMMARY of INCOME and EXPENSES (Cash Basis)

Income			Expenses		
Description	Page	Total	Description	Page	Total
Sales of crops, total	40	$	Farm expenses, total	40	$
Other farm income, total	40		Hog purchases, total	10	
Sale of raised livestock, total	40		Beef cattle purchases, total	10	
Sale of purchased livestock, total	40		Dairy cattle purchases, total	11	
Cooperative distributions paid out, total	41		Other livestock purchases, total	11	
Total cash farm income		$	Total cash farm expenses		$
Net farm income, cash basis (total cash farm income minus total cash farm expenses)					$

Round entries to the nearest whole dollar.

ENDING INVENTORIES of CROPS and SUPPLIES

INVENTORY of CROPS

End of Year

	Quantity	Price	Total Value
Corn		$	$
Corn			
Corn			
Soybeans			
Soybeans			
Silage			
Wheat			
Oats			
Other grains			
Alfalfa hay			
Other hay			
Straw			
Growing crops			
Total crops			$
Minus: Total value from end of previous year			
Equals: Change in inventory value of crops			$

INVENTORY OF SUPPLIES

End of Year

	Quantity	Price	Total Value
Seed		$	$
Fertilizer			
Pesticides			
Fuel			
Commercial feed			
Livestock products			
Prepaid expenses			
Total supplies			$
Minus: Total value from end of previous year			
Equals: Change in inventory value of supplies			$

Round values to the nearest whole dollar. Changes may be negative.

ENDING INVENTORIES of LIVESTOCK

HOG INVENTORY

End of Year

Type	No.	Total Weight	Price	Total Value
Sows			$	$
Gilts				
Boars				
Total breeding stock			xx	
Market hogs				
Market hogs				
Market hogs				
Feeder pigs				
Nursery pigs				
Total market hogs			xx	
Total market plus breeding			xx	
Minus: totals from end of previous year			xx	
Equals: change in inventory value			xx	

OTHER LIVESTOCK INVENTORY

End of Year

Type	No.	Total Weight	Price	Total Value
			$	$
Total breeding			xx	
Total market			xx	
Total market plus breeding			xx	
Minus: totals from end of previous year			xx	
Equals: change in inventory value			xx	

BEEF CATTLE INVENTORY

End of Year

Type	No.	Total Weight	Price	Total Value
Cows			$	$
Bred heifers				
Heifer calves				
Bulls				
Total breeding stock			xx	
Calves				
Yearlings				
Total market cattle			xx	
Total breeding plus market cattle			xx	
Minus: totals from end of previous year			xx	
Equals: change in inventory value			xx	

DAIRY CATTLE INVENTORY

End of Year

Type	No.	Total Weight	Price	Total Value
Cows			$	$
Bred heifers				
Heifer calves				
Bulls				
Total breeding stock			xx	
Calves				
Yearlings				
Total market cattle			xx	
Total breeding plus market cattle			xx	
Minus: totals from end of previous year			xx	
Equals: change in inventory value			xx	

Total change in inventory value of livestock, all species	

Round values to the nearest whole dollar. Changes may be negative.

ENDING INVENTORY OF LOANS and CREDITS

	End of Year				Verification (optional)			
	a	b	c	d	e	f	g	h
Description or Source	Balance Owed	Principal Due in Next 12 Months	Remaining Principal (a - b)	Accrued Interest	Balance Owed, End of Previous Year	New $ Borrowed (page 38)	Principal Paid (page 38)	Apparent Balance Owed (e + f - g)
	$	$	$	$			$	$
Totals	$	$	$	$	$	$	$	$
Minus: accrued interest from end of previous year				$	Round values to the nearest whole dollar.			
Equals: change in accrued interest				$	Totals for column (a) and column (h) should be the same.			

ENDING INVENTORY of ACCOUNTS PAYABLE and RECEIVABLE

Accounts Payable		Accounts Receivable	
Description	Total Value	Description	Total Value
Farm taxes	$		$
Total account payable	$	Total accounts receivable	$
Minus: total account payable from end of previous year	$	Minus: total accounts receivable from end of previous year	$
Equals: change in accounts payable	$	Equals: change in accounts receivable	$

NET WORTH STATEMENT — END OF YEAR

ASSETS (What We Own)			LIABILITIES (What We Owe)		
CURRENT ASSETS	Page		CURRENT LIABILITIES	Page	
CASH (for current use)			Principal due in next 12 months	44	
Checking					
Savings and time deposits			Total accrued interest	44	
Inventory value of crops	42		Total accounts payable	44	
Inventory value of supplies	42				
Inventory value of hogs—market	43				
Inventory value of beef cattle—market	43				
Inventory value of dairy cattle—market	43				
Inventory value of other livestock—market	43				
Accounts receivable	44				
(1) TOTAL CURRENT ASSETS		$	(4) TOTAL CURRENT LIABILITIES		$
LONG-TERM ASSETS			LONG-TERM LIABILITIES		
Inventory value of hogs—breeding	43		Remaining principal	44	
Inventory value of beef cattle—breeding	43				
Inventory value of dairy cattle—breeding	43				
Inventory value of other livestock—breeding	43				
Remaining value of machinery, equipment	41				
Remaining value of vehicles (farm share)	41				
Remaining value of other intermediate assets	41				
Accumulated value of cooperative distributions	41				
Remaining value of buildings	41				
Remaining value of improvements	41				
Remaining value of land	41				
Remaining value of other long-term assets	41				
(2) TOTAL LONG-TERM ASSETS		$	(5) TOTAL LONG-TERM LIABILITIES		$
(3) TOTAL ASSETS (line 1 + line 2)		$	(6) TOTAL LIABILITIES (line 4 + line 5)		$
Round all values to the nearest whole dollar.			(7) NET WORTH (line 3 minus line 6)		$

SUMMARY of INCOME and EXPENSES (ACCRUAL BASIS)

Income	Page	Amount	Expenses	Page	Amount
Total cash farm income	41	$	Total cash farm expenses	41	$
Change in inventory value of crops	42		Depreciation this year	41	
Change in inventory value of livestock	43		Change in accrued interest	44	
Change in accounts receivable	44		Change in accounts payable	44	
Value of farm products consumed at home	7		Minus change in inventory value of supplies	42	
Total farm income		$	Total farm expenses		$
Net farm income, accrual basis (total farm income minus total farm expenses)					$

CROP RECORD

Kind of Crop	Own or Rent O, R	a	b	c	d	e	f	g
		Production			Value			
		Acres	Yield per Acre	Total Bushels a x b	Average Price for the Year	Total Value c x d	Value of Landlord's Share	Operator's Share e - f
Corn					$	$	$	$
Corn								
Corn								
Soybeans								
Soybeans								
Silage								
Wheat								
Oats								
Other grains								
Alfalfa								
Hay—other								
Straw								
Pasture (rental value)								
Total crop acres		A.	Total value of crops raised			$	$	$
Timber								
Waste								
Farmstead								
Roads								
Total farm acres		A.						

Round values to the nearest whole dollar.

SUMMARY of LIVESTOCK INCREASE

1 Change in inventory value of livestock, all species (p. 43)	$
2 Sales of raised livestock (p. 40)	$
3 Sale of purchased livestock (p. 40)	$
4 Value of livestock consumed at home (p. 7)	$
5 Total credits (lines 1 + 2 + 3 + 4)	$
6 Livestock purchases, sum of all species (p. 41)	$
7 Net livestock increase (line 5 - line 6)	$

SUMMARY of FEED USED

8 Inventory value of crops from previous year (p. 42)	$
9 Total feed purchased (p. 40)	$
10 Operator's share of total crops raised (p.46, column g)	$
11 Total supply (lines 8 + 9 + 10)	$
12 Crops sold (p. 40)	$
13 Ending inventory value of crops (p. 42)	$
14 Value of own crops used for seed	$
15 Total used, not fed (lines 12 + 13 + 14)	$
16 Value of feed fed (line 11 - line 15)	$

VALUE of FARM PRODUCTION

17 Total farm income (p. 44)	$
18 Livestock purchases (line 6)	$
19 Feed purchases (line 9)	$
20 Value of farm production (line 17 - line 18 - line 19)	$

HOW EFFICIENT IS OUR FARMING?

	SIZE of BUSINESS	Our Farm	Comparison *
1	Total farm income (page 47, line 17)	$	$
2	Value of farm production (page 47, line 20)	$	$
3	Total crop acres (page 46, column 1)	A.	A.
4	Value of ending livestock inventory, market and breeding (page 45, total for all species)	$	$
5	Value of total assets (page 45)	$	$
6	Net worth (page 45)	$	$
7	Months of labor (operator and family ______ + hired ______)	Mos.	Mos.
	CROP YIELDS and PRICES		
8	Corn yield (page 46, total production ÷ no. of acres)	Bu./Acre	Bu./Acre
9	Corn price (page 1, total sales ÷ total bushels sold)	$/Bu.	$/Bu.
10	Soybean yield (page 46, total production ÷ no. of acres)	Bu./Acre	Bu./Acre
11	Soybean price (page 1, total sales ÷ total bushels sold)	$/Bu.	$/Bu.
12	Total value of crops per acre (page 46, column e, ÷ line 3)	$	$
	LIVESTOCK RETURNS and PRICES		
13	Net livestock increase (page 47, line 7)	$	$
14	Value of feed fed (page 47, line 16)	$	$
15	Livestock income over feed costs (line 13 minus line 14)	$	$
16	Livestock returns per $100 feed fed (line 13 ÷ line 14) x 100	$	$
17	Hog sale price (page 3, total sales, raised and purchased, ÷ total weight sold)	$	$
18	Beef cattle sale price (page 4, total sales, raised and purchased, ÷ total weight sold)	$	$
19	Milk sale price (page 7, total dairy sales ÷ total quantity sold)	$	$
	USE of LABOR		
20	Crop acres per person (line 3 ÷ line 7) x 12	A.	A.
21	Value of farm production per person (line 2 ÷ line 7) x 12	$	$
	PROFITABILITY		
22	Net farm income, accrual (page 44)	$	$
23	Unpaid labor value (line 7, operator + family mos. x $_____/month)	$	$
24	Equity cost (line 6 x _______% return)	$	$
25	Return to management (line 22 - line 23 - line 24)	$	$
	FINANCIAL RATIOS		
26	Interest expense (page 38, total interest + page 44, change in accrued interest)	$	$
27	Debt to asset ratio (page 45, total liabilities ÷ total assets)	%	%
28	Current ratio (page 45, current assets ÷ current liabilities)	%	%
29	Asset turnover ratio (line 2 ÷ line 5)	%	%
30	Net farm income ratio (line 22 ÷ line 2)	%	%
31	Return on assets (line 22 - line 23 + line 26) ÷ line 5	%	%
32	Return on equity (line 22 - line 23) ÷ line 6	%	%
33	Value of farm production per $ of expense (line 2 ÷ (line 2 - line 22))	$	$

* Comparison can be with last year's values, averages for similar farms, or long-term goals.